Not a Chance!

Story by Jill McDougall

Illustrations by Jake Hill

Contents

Chapter 1

What Bad Luck!

"Check this out, Cooper!"

My best friend, Sami, pushed off down the skateboard ramp, her dark curls bouncing. When she reached the bottom, she spun around in a circle.

"Awesome!" I told her.

It was Saturday morning, and I was at the skatepark watching the skaters show off their moves. If you're wondering why I wasn't skating too, well ... it's my own fault. A few weeks ago, I left my skateboard in the driveway where Dad parks his ten-tonne truck.

You can probably guess the rest!

"What bad luck!" Sami had said, when I told her my skateboard was smashed.

But it wasn't bad luck at all. In fact, it was a simple sum: one small skateboard plus one massive truck equals one major disaster.

So now I spent my weekends with nothing to do, and I was bored, bored, *bored*.

"It's your turn, Cooper!" Sami said kindly. She pushed her board towards me, but instead of grabbing it, I let out a gasp and pointed at the road.

A line of trucks was snaking into town. Some were decorated with colourful fruit, and others were painted with laughing clowns.

"The carnival is here!" I yelled.

"Hooray!" cried Sami, high-fiving me with excitement.

"I've saved some pocket money," I told Sami eagerly. "We can go to the carnival tomorrow."

Suddenly, the weekend was looking much better.

Chapter 2

Carnival King

The next morning, as soon as my jobs around the house were done, Sami and I headed to the carnival. The sound of lively pop music greeted us as we hurried through the carnival gates. I caught a glimpse of the Ferris wheel spinning lazily against the blue sky. Overhead, a rollercoaster rattled along its tracks.

Sami tugged at my sleeve, her eyes gleaming with mischief. “Let’s go and find the scariest ride!” she said.

I nodded, but as I glanced around, my brow furrowed. Although it was almost noon, the carnival grounds were strangely empty.

“Where is everyone?” I asked.

Just then, we heard a wild shriek and the buzz of excited chatter. We hurried towards the sounds until we reached a row of carnival games. I slid to a halt and stared at the sight before me. People were pushing and shoving one another, desperate to play one of the games.

Sami and I gazed at one another, astonished. "What's going on?" Sami asked.

At that moment, someone at the Can Toss game let out a yell. I looked across to see Lonnie Finkle pumping his fist in the air. Lonnie was our school's most annoying pest (as voted by me).

"I'm the greatest!" Lonnie was yelling. "I've won a purple token!"

Lonnie's friends let out a cheer as if he had won the lottery.

"What's so special about a purple token?" Sami asked quietly. "I'd rather win a stuffed bear."

Sami and I wandered past a booth lit up with twinkling lights.

"Hurry! Hurry!" boomed a voice through a megaphone. "Time is running out for today's exciting competition!"

I peered inside the booth to see a bearded man standing beside a long golden box. He wore a tall black hat, and the nametag on his waistcoat said "Carnival King".

"Time is running out!" he repeated, quirking an eyebrow at me.

"What for?" I asked.

"To win the greatest carnival prize ever," said Carnival King. With a sweeping gesture, he pointed to the long golden box behind him. "Behold!"

I took a step forward and let out a gasp. Inside the box was the *best* skateboard in the whole entire world. I gazed longingly at its red shiny wheels and smooth, curved deck.

Sami nudged me, eyes shining. "You have to win that skateboard!" she said excitedly. "Then we can skate together again."

Carnival King flashed me a wide smile. "All you need is three purple tokens," he said.

He waved an arm at the carnival games. "Play any game to win. The first player with three purple tokens walks away with the grand prize."

"Wow!" said Sami.

"You'll hear a siren blast when the prize has been won," explained Carnival King. He scratched his chin, as if remembering something important, then said, "One more thing! You'll be given a stamp on your wrist when you win a token. We don't want players combining their tokens."

At that moment, a cry went up from the Duck Pond game behind us. It was Lonnie Finkle. "Yippee!" he yelled. "I've won my second token!"

Carnival King fixed me with an urgent look. "You'd better get started," he said. "The contest closes in one hour."

My shoulders slumped. "I have no chance against Lonnie Finkle," I murmured glumly. "He's an expert at all sorts of games."

"Don't give up, Cooper," whispered Sami. "I have a plan."

Chapter 3

Fair Game?

Sami and I gazed at the row of brightly coloured carnival games. All around us, people hurried from one game to the next like ants at a picnic.

"What's your plan?" I asked.

Sami flashed me a cheeky grin. "Look around," she said. "People are trying to win tokens by playing random games."

I bunched my eyebrows, puzzled. "Isn't that the idea? The more games you play, the more chances you have of winning a purple token."

“All carnival games are different,” said Sami. “Some give you hardly any chance of winning and others are much fairer.” Her grey eyes narrowed. “You need to play the fairer games, right?”

“Right,” I said, feeling more cheerful. I glanced at my watch and let out a gasp. “Only fifty minutes until the contest closes!” I exclaimed. “Let’s start checking out the games.”

We stopped at a stall where a sign with flashing lights said “Barrel of Luck”. The game looked simple enough. Players were blindfolded and led to a clear plastic barrel full of tokens. All they had to do was reach into the barrel and pull one out.

"This game looks fair," I muttered to Sami. "There must be a hundred tokens in that barrel."

"Look more closely," said Sami. "Most of the tokens are yellow and a few are blue. I can only see one purple one."

I heaved a sigh. "So, my chance of winning a purple one would be … one in a hundred. I'm much more likely to pull out a …"

"… yellow one!" moaned a player, as she took off her blindfold.

We left the Barrel of Luck and headed towards a game called Ring Toss. Here, dozens of glass bottles were lined up in rows, and every bottle had a purple token on top.

Players were given plastic rings to throw at the bottles. If a ring landed around the neck of a bottle, the player won the token.

"This looks easy," I whispered to Sami. "There are so many bottles, the likelihood of missing is practically zero."

Sami frowned. "There must be a catch," she said, gazing at the bottles. "Ring Toss might look simple, but …"

I didn't wait to hear the rest. Time was running out, and I needed to start winning tokens before Lonnie Finkle beat me to the prize.

I paid the money and the game operator handed me a bucket of plastic rings.

My first throw was a really good shot. The ring headed straight for the top of the bottle, but then ... *ping!* It bounced off the glass and landed at my feet. I threw the next ring and the next, but the same thing happened every time.

Soon, my bucket was empty.

"What bad luck!" I moaned, collecting up the plastic rings.

"It's not bad luck," said Sami. "See how the bottles are tightly packed? That's so the ring will hit two bottles at once and bounce off."

I tapped one of the rings with my knuckles. It was made from super-hard plastic – just the kind of material that would bounce off glass. Suddenly, the truth sank in. It would be almost impossible to win at Ring Toss! We needed to move on to another game.

My heart took a dive. Time was running out, and I was no closer to winning that skateboard.

Chapter 4

Lucky Break

"Don't give up, Cooper," said Sami. "We still have thirty minutes before the contest closes!"

But as we moved from game to game, my spirits sank lower. All around us, players were desperately trying to win purple tokens, but no one seemed to have won any.

No one, that is, except Lonnie Finkle.

At the end of the alley, I spotted a game stall with a board of balloons in different colours – red, blue, yellow and purple. A sign said "Balloon Pop".

“I know this game,” I told Sami excitedly. “I played it last year and won heaps of prizes.”

The booth appeared deserted until a ginger-haired man emerged from the shadows and regarded us hopefully. “Try your luck at Balloon Pop!” he said, thrusting some darts towards me. “If you burst a purple balloon, you win a purple token.”

“You get three darts!” said Sami breathlessly. “That means you could win three tokens.”

I checked out the balloons to make sure they were properly inflated. A balloon without plenty of air pressure would be hard to burst.

They looked fine, so I paid the man and got ready to take my first shot. At that moment, Lonnie Finkle and his friends arrived. Lonnie watched me with his hand on his hip and a smirk on his face.

I wiped the sweat from my forehead and lined up my first dart. I was aiming for a fat purple balloon right at the top.

“Hit it hard and fast,” urged Sami.

I took a deep breath and threw as hard as I could. The dart flew out of my hand like a missile. Someone in the crowd let out a low whistle as the dart headed straight for the balloon.

Pop! The purple balloon burst!

"Good shot!" cried Sami.

There were low murmurs from Lonnie and his friends as the carnival attendant stamped my wrist and handed me a purple token.

I got ready for my second throw, placing my feet carefully on the painted line. Raising the dart, I closed my eyes and imagined it hitting its target. Then I opened my eyes and threw.

Pop! Another purple balloon hung in shreds.

Sami gripped my shoulder hard. "One more, Cooper," she said. "Come on, you can do it!"

"I bet you can't!" said someone behind me.

It was Lonnie Finkle, standing so close I could feel his hot breath on my neck.

"Move back!" I told him firmly. "Give me room."

A muscle twitched in my cheek as I lined up my last dart. I took aim, but just as I was about to throw, a loud barking cough broke the silence. My hand shook, and the dart launched itself into the air and wobbled limply towards the balloon.

I stared in dismay as it fell short and dropped to the ground.

Lonnie caught my eye and gave me a smirk. Sami stared back at him angrily. Everyone knew Lonnie had coughed on purpose.

"It's okay," I told Sami, as I fished for coins in my pocket. "I'll play again."

But Sami shook her head sadly and pointed to a sign that said *ONE GAME PER PLAYER.*

Disappointment surged through me.

"I was so close to winning," I mumbled.

"Don't give up now," Sami said. "We just need to find one last game with good odds."

"Balloon Pop was the last game," I reminded her. "We've seen all the others."

I glanced at my watch in dismay. The contest closed in ten minutes, and my chance of winning that skateboard was zero.

Chapter 5
Fifty-Fifty

Sami and I were passing Carnival King's stall when we noticed something interesting. A woman wearing a green jacket was putting together a colourful wooden wheel.

"Look at that," said Sami in surprise. "Here's something you haven't tried."

The woman noticed us watching her. "Do you want to play the Monster Spinner?" she asked. "It's all ready to go."

I studied the wheel closely. It was divided into twelve triangles like a pizza, and every second triangle was painted purple.

"How … how do you win?" I asked.

"Just spin the wheel," the woman replied. "If the pointer lands on purple, you win a purple token."

"That means you have a fifty-fifty chance of winning!" squeaked Sami.

The woman gently tugged at the wheel and the colours began to spin … purple, red, purple, blue, purple, yellow … until they became a blur.

The wheel made a *tick, tick, tick* sound as the plastic pointer clicked from one triangle to the next. Finally, it came to rest on a purple triangle.

Sami grabbed my arm. "Quick, have a go," she said. "We still have a few minutes left!"

My heart raced as I pulled some coins out of my pocket. I was about to pay when I felt a push on my back. I lost my balance and landed on my knees in the dirt. As I got to my feet, I saw Lonnie Finkle looking pleased with himself. He was holding out some coins to the woman in the green jacket.

"Hurry up," he said. "I want to play your spinner. I only need one more token to win!"

The woman stared at Lonnie with a look of disgust.

"I saw you push that boy over," she said, frowning. "Now you can wait your turn."

Then she pointed to me. "Off you go."

Before I knew it, I was tugging on the giant wheel and watching the colours spin. Finally, it came to rest on … a red triangle.

What a shame!

"Try again," said the woman, with a wink. "Every player gets two spins."

Well, you can probably guess what happened. The spinner landed on a purple triangle!

Next minute, I was racing towards Carnival King with three stamps on my wrist and clutching three purple tokens.

Luckily, Carnival King saw me coming, because he was just about to close the contest.

"Well done!" he said, as he handed over the skateboard. "You've earned it!"

I was grinning so hard I felt like my face might burst.

ster Spinner
GAME
LAYER

Later, Sami and I tested out the new skateboard in my driveway. It was awesome!

"I'm going to look after my skateboard this time," I promised Sami. "I'll keep it under my bed."

"Not in the driveway?" teased Sami.

"Not a chance," I told her. "Not a chance!"